The Science Of Christmas: A Miscellany

Tim Sandle

DEDICATION

This book is dedicated to Jenny Sandle and Jake Roberts

CONTENTS

INTRODUCTION

Welcome to the *Science Behind Christmas*. This short book is designed to take a whimsical look, from then scientific perspective, at some of the customs and myths associated with the Christmas holiday season. This is cold, hard science, for the myths and traditions are explained and kept fresh. In many cases, the scientific perspective is just for fun.

The book, at times, is designed as reference guide, giving you footnotes in places should you wish to read further (or check the facts stated). There are some illustrations to, either to put points into perspective or to color the points being made. There is also the odd puzzle or two!

Importantly, the book is a miscellany, it is not designed to be a comprehensive account of the history of Christmas traditions or an explanation of all scientific related Christmas matters. Rather it is a mixture, medley, and assortment of facts and trivia.

For those interested in the science behind other traditions, there is a bonus section relating to the science behind Halloween.

These look like the types of ornaments a budding scientist might have over the holiday season!

The Science Of Christmas: A Miscellany

1 SANTA

Picture 1: Santa on his sleigh, moving rather slowly given what he needs to achieve!

Synonymous with Christmas, or at least its secular side, is Santa Claus (or Father Christmas or St. Nicholas, depending upon the tradition).

The story of Santa Claus begins with Nicholas, who was born during the third century in the village of Patara, an area now on the southern coast of Turkey. According to legend, his wealthy parents, who raised him to be a devout Christian, died in an epidemic while Nicholas was still young. Following the Christian teaching of giving the money to the poor, Nicholas used his inheritance to assist the needy, the sick, and the suffering. It is from this that the idea of a person giving gifts at a certain time of year comes from. The anniversary of Nicholas' death became a day of celebration, St. Nicholas Day, December 6th (December 19 on the Julian Calendar).

The name Santa Claus evolved from Nicholas' Dutch nickname, Sinter Klaas, a shortened form of Sint Nikolaas (Dutch for Saint Nicholas). In 1804, John Pintard, a member of the New York

Historical Society, distributed woodcuts of St. Nicholas at the society's annual meeting. The background of the engraving contains now-familiar Santa images including stockings filled with toys and fruit hung over a fireplace.

As to how the figure became popular again and associated with Christmas probably dates to 1809, at least in North America. Here Washington Irving helped to popularize the Sinter Klaas stories when he referred to St. Nicholas as the patron saint of New York in his book, The History of New York. As his prominence grew, Sinter Klaas was described as everything from a "rascal" with a blue three-cornered hat, red waistcoat, and yellow stockings to a man wearing a broad-brimmed hat and a "huge pair of Flemish trunk hose."

1840s, newspapers were creating separate sections for holiday advertisements, which often featured images of the newly-popular Santa Claus.

In parallel, in parts of Europe, Christkind or Kris Kringle was believed to deliver presents to well-behaved Swiss and German children. Meaning "Christ child," Christkind is an angel-like figure often accompanied by St. Nicholas on his holiday missions.

Similar figures appear elsewhere:

- In Scandinavia, a jolly elf named Jultomten was thought to deliver gifts in a sleigh drawn by goats.
- British legend explains that Father Christmas visits each home on Christmas Eve to fill children's stockings with holiday treats.
- Pere Noel is responsible for filling the shoes of French children.
- In Russia, it is believed that an elderly woman named Babouschka purposely gave the wise men wrong directions to Bethlehem so that they couldn't find Jesus. To this day, on January 5, Babouschka visits Russian children leaving gifts at their bedsides.
- In Italy, a similar story exists about a woman called La

Befana, a kindly witch who rides a broomstick down the chimneys of Italian homes to deliver toys into the stockings of lucky children.

Overtime, the association between Santa / St. Nicholas / Father Christmas became steeped into many cultures and traditions. How can one man, even with mystical powers, visit so many children in such a short space of time?

Let's begin with this dilemma, and one that all children must grapple with at some time or other: how fast would Santa Claus need to travel to deliver presents to the children of the world?

In answering this, first, an assumption needs to be made. This is that Santa has to travel 510,000,000 kilometers on Christmas Eve, and that he has around 32 hours to do it (remember he will be able to take advantage of different time zones). This is due to the rotation of Earth he'll have a good 31 hours to deliver presents - rather than just 24 - assuming he's flying East to West to match Earth's rotation.

To achieve this, Santa will need to travel at 10,703,437.5 kilometers per hour, or about 1,800 miles per second, all night (assuming he never stops and comes equipped with some sort of sleigh-mounted present-launcher will be required to shoot gifts down chimneys while moving)[1].

---

[1] See: http://www.telegraph.co.uk/topics/christmas/8188997/The-science-of-Christmas-Santa-Claus-his-sleigh-and-presents.html

Or, to put it another way via this Infographic (**Picture 2**):

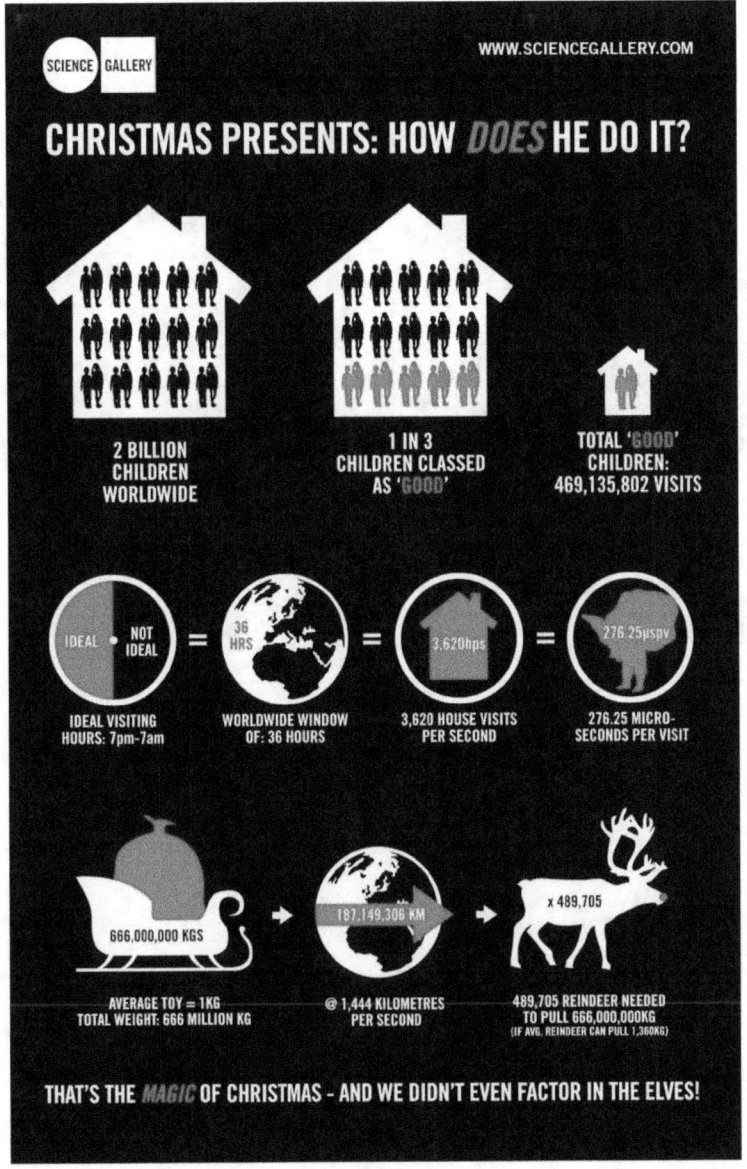

How about the weight of the sleigh? If Santa was carrying toy robots for children and he had 700 million of them, then his sleigh would weigh 1,232,300 metric tons and this would need in the region of three million reindeer to pull it[2].

Picture 3: Maybe Santa uses a rocket powered sleigh?

---

[2] See: http://news.ncsu.edu/releases/wmssilverbergsantasleigh/

Or more likely, as the infographic shows, physics may have the answer:

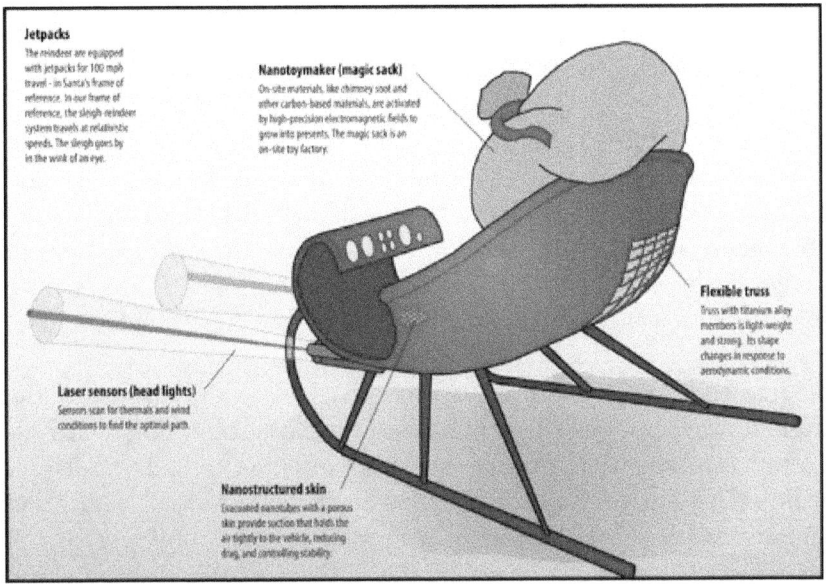

Picture 4: The physics of Santa's sleigh.

Here's **an alternative take** on the Santa travel conundrum:

1) No known species of reindeer can fly. But it is estimated that there are 300,000 species of living organisms yet to be classified, and while most of these are insects and germs, this does not COMPLETELY rule out flying reindeer, which only Santa has ever seen.

2) There are 2 billion children (persons under 18) in the world. But since Santa doesn't (appear) to handle the Muslim, Hindu, Jewish and Buddhist children, that reduces the workload to 15% of the total - 378 million according to the Population Reference Bureau. At an average (census) rate of 3.5 children per household, that's 91.8 million homes. One presumes there's at least one good child in each.

3) Santa has 31 hours of Christmas to work with, thanks to the different time zones and the rotation of the earth, assuming he travels east to west (which seems logical). This works out to 822.6 visits per second. This is to say that for each Christian household with good children, Santa has 1/1000th of a second to park, hop out of the sleigh, jump down the chimney, fill the stockings, distribute the remaining presents under the tree, eat whatever snacks have been left, get back up the chimney, get back into the sleigh and move on to the next house.

Assuming that each of these 91.8 million stops are evenly distributed around the earth (which, of course, we know to be false but for the purposes of our calculations we will accept), we are now talking about .78 miles per household, a total trip of 75-1/2 million miles, not counting stops to do what most of us must do at least once every 31 hours, plus feeding, etc. This means that Santa's sleigh is moving at 650 miles per second, 3,000 times the speed of sound. For purposes of comparison, the fastest man-made vehicle, the Ulysses space probe, moves at a poky 27.4 miles per second - a conventional reindeer can run, tops, 15 miles per hour.

4) The payload on the sleigh adds another interesting element. Assuming that each child gets nothing more than a medium-sized lego set (2 pounds), the sleigh is carrying 321,300 tons, not counting Santa, who is invariably described as overweight. On land,

conventional reindeer can pull no more than 300 pounds. Even granting that "flying reindeer" (see point #1) could pull ten times the normal amount, we cannot do the job with eight, or even nine. We need 214,200 reindeer. This increases the payload - not even counting the weight of the sleigh - to 353,430 tons. Again, for comparison - this is four times the weight of the Queen Elizabeth.

5) 353,000 tons traveling at 650 miles per second creates enormous air resistance - this will heat the reindeer up in the same fashion as spacecraft re-entering the earth's atmosphere. The lead pair of reindeer will absorb 14.3 QUINTILLION joules of energy, per second,.each. In short, they will burst into flame almost instantaneously, exposing the reindeer behind them, and create deafening sonic booms in their wake. The entire reindeer team will be vaporized within 4.26 thousandths of a second. Santa, meanwhile, will be subjected to centrifugal forces 17,500.06 times greater than gravity.

A 250-pound Santa (which seems ludicrously slim) would be pinned to the back of his sleigh by 4,315,015 pounds of force.

In conclusion - If Santa ever DID deliver presents on Christmas Eve, he's dead now.

Finally, with another tradition, many people leave food and drink out for Santa. This means he'll eat and drink quite a bit:

- Based on population figures, Santa will deliver gifts to 1.6 billion children.
- This requires visiting 5,556 homes a second and eating 150 billion calories in milk and mince pies (If Santa eats milk and mince pies at each of these addresses, he will consume a total of 150 billion calories in just one night - 60,000 times his daily recommended intake).
    - Assuming each of the 640 homes he visits gives him a 200ml glass of semi-skimmed milk and one mince pie, Santa will have drunk almost 130 million liters of milk by the time he's finished his deliveries - enough to fill over 50 Olympic swimming pools.
    - He will also have eaten nearly 40,000 metric tons worth of mince pies.
    - To work off this extra weight, Santa would need to walk 1.3 billion miles, which is 54,000 times around the circumference of the Earth.
- With an average of 2.5 children per household, Santa will need to make 640 million stops on Christmas Eve.
- Each child needs 80cm of wrapping, which would stretch 1.5 million miles.
    - Assuming each of Santa's elves can wrap a present in exactly 10 seconds and at that speed wrapping two presents for every child would require 3,000 elves to work eight hours every day for an entire year to get the job done.
    - Furthermore, if Santa gave two presents to each child the elves would need to use an average of 80cm of wrapping paper for each gift – which works out at a staggering 1.6 million miles of paper.
- The total number of presents would set Santa back £279.27 billion (in U.K. currency).
- Prior to the big night, Santa needs to store all these presents in a warehouse. Assuming each present average out at 0.008 m^3, the warehouse would need to cover the same space as 240,000 double-decker buses.

Santa would also need a big database to deal with all the letters from children. Based on the world population data for under-14s, taken from the U.S. Global Census, Santa would receive an estimated 1,843,868,987 letters each year.

If each Christmas letter takes up an estimated 500KB when uploaded as a scanned copy, it would take up 859GB of data each year. Letters from India's children would take up the most room with 338MB. China comes in second taking up 225MB and Nigeria sits in third place taking up 72MB. So, Santa would need a super computer.

Interim - Christmas math problems #1

Want to try out some Christmas math problems before moving onto the next chapter?

1. If candy canes cost .89 a dozen, how much would it cost to buy candy canes for a school with 400 students?

2. Macy's has hired 400 store Santas. If each Santa sees 125 children a day for 30 days, how many children are seen by

3. One group of carolers goes to every 6th house in a neighborhood, another goes to every 8th house. At which house will they first meet?

4. Mr. Green is putting lights around 8 windows; each window is 3 and 1/2 feet wide and 5 feet long. How many feet of lights does he need?

5. 10 elves each made 10 xylophones. Each xylophone had 10 keys. They did this for 10 days. How many keys were made by elves? Show in exponential form and solve.

6. Each batch of 48 cookies that Amy makes takes 20 minutes in the oven. If the oven is on for 3 hours and 55 minutes (15 minutes for preheating), how many cookies did Amy make?

7. Beth is making gingerbread men. She uses 2 raisins for eyes and 3 raisins for buttons for each gingerbread man. She buys 4 boxes of raisins, each with 120 raisins in it. How many dozens of gingerbread men can she make?

8. Brian wants a 7 and 1/2 foot Christmas tree. Which of these trees is a better buy?

A.) All trees $42
B.) $6 a foot
C.) First 4 feet $22; each additional foot $6

9. K-Mart is open Monday through Saturday from 8:00 AM to midnight and Sunday from 10:00 AM to 8:00 PM. How many hours is it open in one week?

10. Each large roll of ribbon has 20 yards. Each package takes 3 feet of ribbon. How many packages can be wrapped?

The answers are over the page.

Math puzzle answers:

1. $30.26

2. 1,500,000 children

3. 24th

4. 136 ft. - 17 for each window

5. 10,000; 10

6. 11 batches in 3 hours 40 minutes - 528 cookies

7. She has 480 raisins so she can make 96 gingerbread men - 8 dozen

8. Choice A

9. 106 hours

10. 20 packages

2 REINDEER

Reindeer are associated with Santa, although the most famous - Rudolph -was a later addition. Rudolph, "the most famous reindeer of all," was created over a hundred years after his eight flying counterparts. The originals are: Dasher, Dancer, Prancer, Vixen, Comet, Cupid, Donner, and Blitzen. They first appear in a poem titled "A Visit from St. Nicholas" (commonly called "The Night Before Christmas"). The poem was written in 1823 by Clement C. Moore. Part of the poem reads:

More rapid than eagles his coursers they came,
And he whistled, and shouted, and call'd them by name:
"Now, Dasher! Now, Dancer! Now, Prancer, and Vixen!
"On, Comet! On, Cupid! On, Dunder and Blixem!

The names of two of the reindeer change a little: Dunder (variously spelled Donder and Donner), and Blixem (variously spelled Blitzen and Blixen). The names of Dunder and Blixem derive from Germanic words for thunder and lightning, respectively.

The red-nosed wonder was the creation of Robert L. May, a copywriter at the Montgomery Ward department store. So, what about the most popular of the perennial reindeer? Rudolph's red nose is explained by air friction and his nose getting too hot as he guides Santa's sleigh and the other reindeer to drop off presents to every home on the globe[3].

---

[3] See:
http://www.edp24.co.uk/news/the_secret_to_rudolph_s_red_nose_explained_at_norwich_science_show_1_3141208

Picture 4: A cheeky reindeer.

Red noses

There is actually a more serious scientific explanation. Scientists have found that a reindeer's nose is red because it is richly supplied with red blood cells, which help to protect the animal from getting too cold and to regulate brain temperature.

Linking scientific study into the festive season, scientists have stated that tiny blood cells in the nose of the reindeer are vital for delivering oxygen, controlling inflammation, and regulating temperature.

Reindeer are a species of deer native to Arctic and Subarctic regions, characterized by antlers. In relation to Christmas, reindeer traditionally help to pull the sleigh of Santa Claus and as he delivers Christmas gifts.

For the research, the scientists compared the proportion of blood vessels in the nose of people and compared them to reindeers. It was found that the reindeer has 25% more densely compacted blood vessels compared to a person.

The researchers also found a high density of mucous glands

scattered throughout the reindeer noses, which they say helps to act as a barrier against the cold. Furthermore, infrared thermal images showed that reindeer do indeed have red noses. The research was undertaken by scientists based in the Netherlands and Norway. The news about the temperature regulation of reindeers has been published in the *British Medical Journal* (on-line)[4].

Color changing eyes

There are some other fascinating facts about reindeer. To survive the wintery conditions in the Arctic, scientists have found that one part of an reindeer's eyes changes color. This adaptation increases the sensitivity of the animal's vision, helping it navigate the through the darker months.

The change to the color of the eyes of reindeer is one facet of the animals' ability to adapt to changing conditions in the Arctic, especially the move towards the harsh winter. Other changes that take place include developing thick coats with two layers of fur; the pads on their hooves shrink in winter to give them better traction; and they have the ability to detect ultraviolet light, which helps them see in the months of near-total darkness.

The part of the eye that undergoes a color change is called the *tapetum lucidum*. This is the layer of tissue behind the retina that reflects light and helps an animal to better see in dim light. It's sometimes referred to as the "cat's eye" because it's the layer that causes the eyes of a cat to glow at night when a small amount of light hits the eye.

In most mammals, this eyeshine is golden, and this includes reindeer in summer. However, in the winter a reindeer's *tapetum lucidum* shines blue. The shift to blue increases the scatter of reflected light so it passes to more photoreceptors, thereby allowing the reindeer to make out more images under very dark conditions. At the same time as appearing blue, the reindeer pupils also dilate.

[4] http://group.bmj.com/group/media/latest-news/experts-discover-why-rudolph2019s-nose-is-red

The new research into eye color changes was undertaken by teams based in Norway and England. The findings have been published in the *Proceedings of the Royal Society B*[5]. The paper is titled "Shifting mirrors: adaptive changes in retinal reflections to winter darkness in Arctic reindeer."

---

[5] See: http://rspb.royalsocietypublishing.org/content/280/1773/20132451

Picture 5: Nobel reindeer, in Siberia.

Keeping safe

Finally, reindeer have another interesting ability. Ultra-violet light-emitting power lines may upset reindeer and other animals, a new study has found. However, the animals are very adept at avoiding power cables.

A study suggests that many species of birds and mammals, including reindeer, avoid power lines because they are able to see the ultraviolet (UV) light the cables given off. People cannot see UV light, however many animals can, including reindeer and caribou. These animals use UV light to find plants under the snow.

With power cables, the scientists are of the view that reindeer see power lines not as dim, passive structures but, rather, as lines of flickering light stretching across the terrain.

The research has been published in the journal *Conservation Biology*[6], the article is titled "Ultraviolet Vision and Avoidance of Power Lines in Birds and Mammals."

Interim - a Christmas math puzzle #2

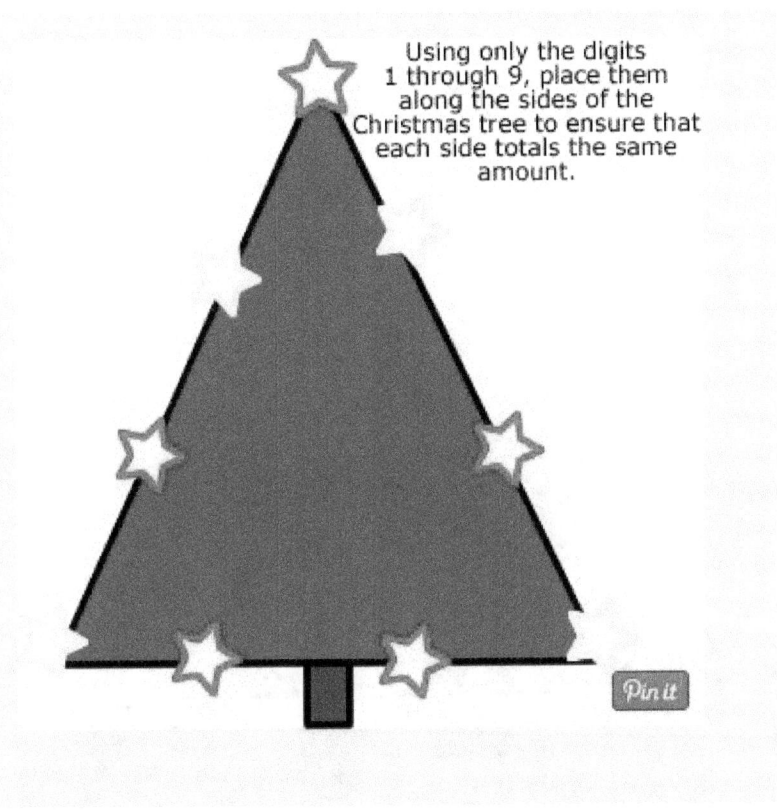

Using only the digits 1 through 9, place them along the sides of the Christmas tree to ensure that each side totals the same amount.

Using only the digits one through 9, place them around the 3 sides of the Christmas tree, making sure that each side totals the same amount. Can you find 2 solutions?

Answers...over the page.

Solution 1:

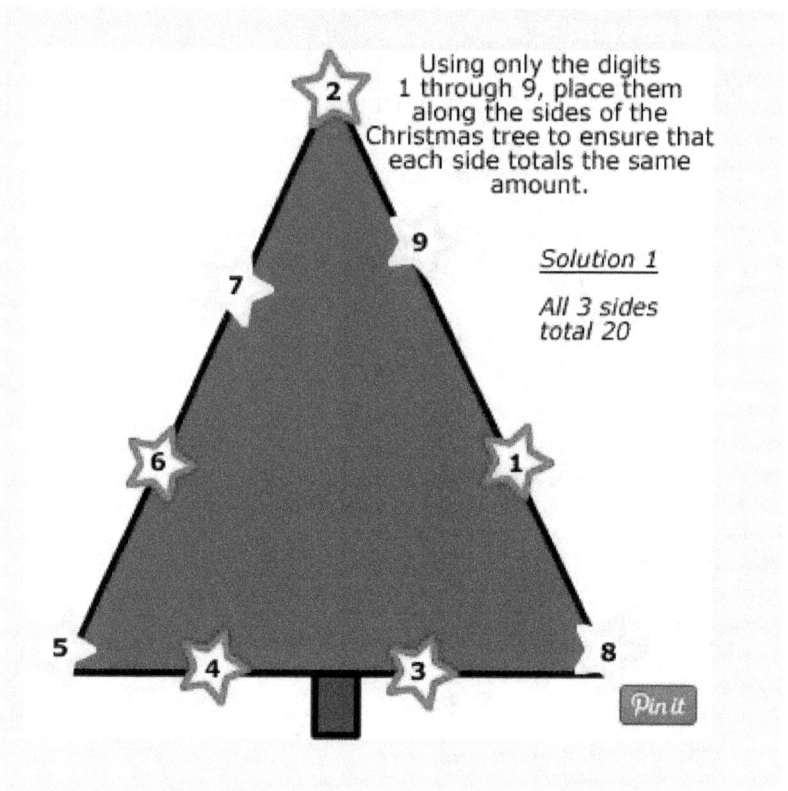

Using only the digits
1 through 9, place them
along the sides of the
Christmas tree to ensure that
each side totals the same
amount.

Solution 1

*All 3 sides
total 20*

All 3 sides total 20 using only the digits 1 through 9.

Solution 2:

All 3 sides total 21 using only the digits 1 through 9.

3 VARIOUS TRADITIONS

There are lots of traditions associated with Christmas, some older than others. In this chapter several of the most popular are presented.

Christmas cards

The custom of sending Christmas cards was started in the UK in 1843 by Sir Henry Cole. He was a civil servant who was very interested in the new 'Public Post Office' and wondered how it could be used more by ordinary (well, middle-class) people.

Sir Henry had the idea of Christmas Cards with his friend John Horsley, who was an artist. They designed the first card and sold them for 1 shilling each. The card had three panels. The outer two panels showed people caring for the poor and in the centre panel was a family having a large Christmas dinner. About 1000 were printed and sold.

The first general postal service started in 1840 when the first 'Penny Post' public postal deliveries began. The new Post Office was able to offer a Penny stamp because new railways were being built. These could carry much more post than the horse and carriage that had been used before. Also, trains could go a lot faster. Cards became even more popular in the UK when they could be posted in an unsealed envelope for one halfpenny - half the price of an ordinary letter.

As printing methods improved, Christmas cards became much more popular and were produced in large numbers from about 1860. In 1870 the cost of sending a post card, and also Christmas cards, dropped to half a penny. This meant even more people were able to send cards. From this a Western hemisphere tradition was born.

As you start to wind down over Christmas does a sudden thought grip you that you may have forgotten to send all of your

Christmas cards? With cards, research suggests that the typical person sends around 150 Christmas cards. Apparently this number relates to the natural size of our social groups and this number of '150' goes back to the typical size of groups established by hunter-gatherer communities in the past[7].

Picture 5: A typical - and comical - Christmas card

Picture 6 (below): A Christmas card infographic

---

[7] See: http://www.newscientist.com/article/mg18024265.900

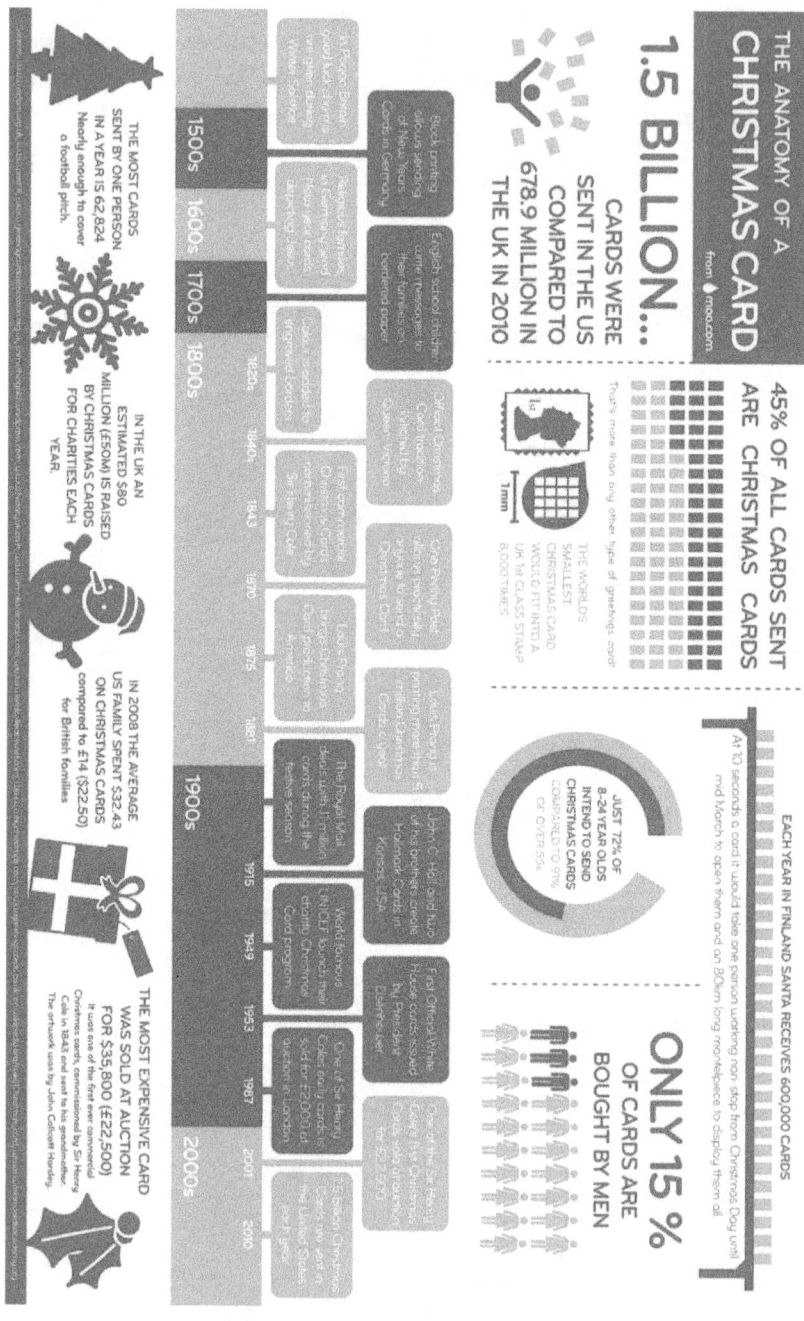

Tinsel

At this time of year many homes, workplaces, shopping malls and the like are decorated with tinsel, a sparkling type of decorative garland material. Why do we do this? Where did the tradition come from?

First off, what is the reason for tinsel? It is said that the use of sparkly material mimics the effect of ice or icicles. To begin with tinsel was a metallic garland for Christmas decoration. Today it is made from plastic (or specially coated paper.)

Picture 7: Tinsel brightens up Christmas

Records indicate that tinsel was invented in Nuremberg around 1610[8]. These early tinsels were made of real silver and required bespoke machines to create the ribbon-like effects that were needed to make the tinsel "sparkle." The inventor of tinsel remains unknown (although there is an amusing German fairytale on the subject

[8] See: http://www.christmascarnivals.com/christmas-history/christmas-history-tinsel.html

involving Kris Kringel and a German aunt.)

However, because it became common to put candles onto Christmas trees, the silver used to make tinsel tarnished easily. Nonetheless., silver tinsels continued to be used around the world until the mid-twentieth century. The next round of tinsel production used a mix of lead and tin. Not only was this toxic (a fact not realized at the time) it was generally too heavy for most Christmas trees.

By the time of the early twentieth entry, manufacturing advances allowed cheap aluminum-based tinsel to be created. This proved a little too costly as the popularity of tinsel grew and the demands of mass production increased. For reasons of scale, modern tinsel is typically made from polyvinyl chloride (PVC) film coated with a metallic finish and sliced into thin strips.

The fact that tinsel has been made out of everything from real silver, to lead to other dangerously flammable materials means that the colorful bands have not always been so safe!

Nutmeg

Now, let's consider overdosing on a little too much festive cheer. No, not alcohol, but nutmeg. The spice is a hallucinogen, although the properties only kick in after large quantities. A sprinkle of nutmeg is bitter-cinnamon delight. But a few teaspoons of the stuff can poison you, because nutmeg contains compounds that carry an intense and rather unpleasant hallucinogenic high. This all happens because nutmeg contains a volatile oil and that oil includes compounds such as myristicin and elemicin[9].

Picture 8: delicious ground nutmeg.

---

[9] See: http://abcnews.go.com/Health/large-doses-nutmeg-hallucinogenic-high/story?id=12347815

Food poisoning

Moving onto Christmas dinner, and here things become a little more serious. It is very important to cook meat properly, especially if you are having turkey. Analysis of bacteria in U.S. turkeys has revealed that high proportions of bacteria found in the birds are "superbugs", resistant to many of the antibiotics used on farms and to treat people. These are not a problem if the food is properly cooked.

Picture 9: A idealized Christmas dinner.

Over-eating

Now, sticking with food, why is it at this time of year we eat too much? Many of us can have enough of one sort of food, but still have an appetite for something else. It lies at the root of the expression "pudding stomach". It appears that this is due to a phenomenon called 'sensory specific satiety' (SSS). Well, it's only once a year...[10]

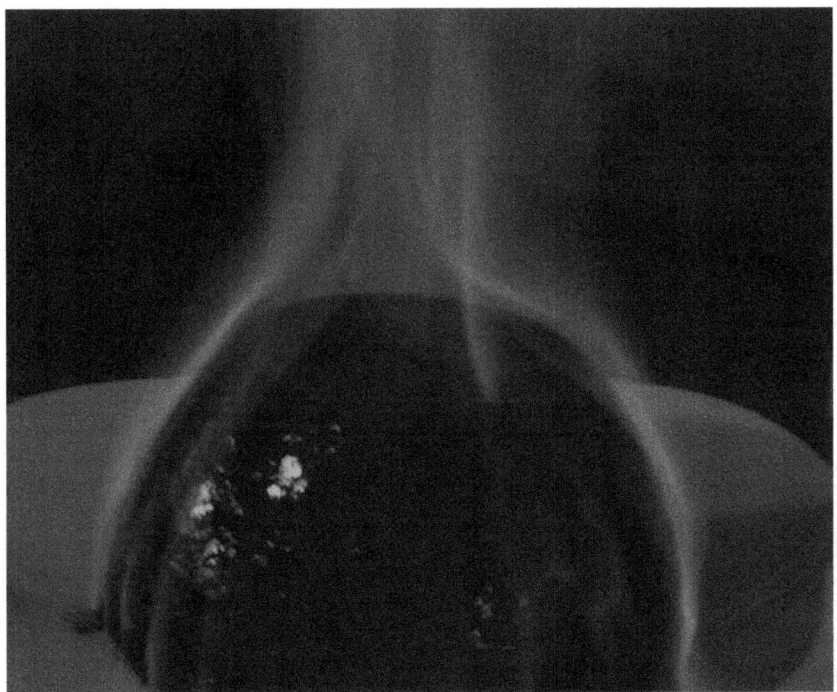

Picture 10: A Christmas pudding on fire.

---

[10] See:
http://www.theguardian.com/lifeandstyle/wordofmouth/2013/dec/17/stomach-christmas-feeling-full-food-and-drink-appetite

Interim - Christmas math problems #3

Some more math problems for the holiday season!

1. If you buy every item in the song, The Twelve Days of Christmas, how many items will you buy?

2. What is the ratio (fraction) of birds to the total number of items?

3. Santa and Mrs. Claus want to take 4 of the elves to see THE NUTCRACKER. How much will it cost if adult tickets are $34.95 and elf tickets (child) are $24.95?

4. Santa needed to draw a map of one of the areas he had to go to deliver gifts. Follow these directions and draw the map for him.

   - Ice Boulevard is parallel to Reindeer Road and Holly Highway.

   - Snowflake Street is diagonal to all 3 and deadends at Holly Highway.

   - Present Pike is perpendicular to Holly, Reindeer, and Ice.

   - Mistletoe Mall is bordered by Reindeer Road, Present Pike and Ice Boulevard.

   DRAW THE MAP AND LABEL EVERYTHING. Use the back of your paper.

5. Mrs. Claus wants to have a nice meal ready when Santa comes home. She wants to make waffles and she knows that she will have to double the recipe since he will be so hungry. Show how much of each ingredient she will need.

- 3/4 stick of butter (melted)
- 1 1/4 cup of flour
- 1/4 cup sugar
- 1/2 tbsp. baking powder
- 1/8 tsp. salt
- 7/8 cup milk
- 1 tsp. vanilla
- 1 egg

6. Mrs. Claus needs to feed the reindeer. She buys 2 pounds of food per day per reindeer. How much food does she need for December? For the whole year?

7. She wants to put lights all around her windows. She has 15 windows all of the same size. Each one is 3 feet long by 2 feet wide. How many strands will she buy if each strand covers 10 feet? How much will she pay if each strand costs $6.99? Don't forget that there is tax of 7%.

Answers are over the page.

Math puzzle answers:

1. (1 x 12) + (2 x 11) + (3 x 10) + (4 x 9) + (5 x 8) + (6 x 7) + (7 x 6) + (8 x 5) + (9 x 4) + (10 x 3) + (11 x 2) + (12 x 1)

   WHICH CAN BE SIMPLIFIED INTO JUST DOUBLING THE FIRST HALF BECAUSE IT REPEATS:

   2((1 x 12) + (2 x 11) + (3 x 10) + (4 x 9) + (5 x 8) + (6 x 7))
   2(12 + 22 + 30 + 36 + 40 + 42)
   2(182) = 364 ITEMS

2. 12 PARTIDGES + 22 TURTLE DOVES + 30 CALLING BIRDS + 36 FRENCH HENS + 42 GEESE + 42 SWANS = 184 BIRDS.

3. 184/364 = 46/91

4. $169.70

5. Many possible solutions. Check to see if each of the 4 criteria is met.

   - 1 1/2 sticks butter
   - 2 1/2 cups flour
   - 1/2 cup sugar
   - 1 tbsp. Baking powder
   - 1/4 tsp. Salt
   - 1 3/4 cups milk
   - 2 tsp. Vanilla
   - 2 eggs

6. 558 pounds for December (including Rudolph)
   6570 pounds for the year.

7. 15 strands; $112.19 with tax

4 CRACKERS

Like to pull a cracker? If you're not familiar with this festive activity, crackers are tubes of cardboard, with a popper, containing a toy, motto and a hat. One person holds each end and the person holding the main part of the cracker wins a prize.

Picture 11: The cracker pulling tradition

With this scenario, how many crackers do you need to ensure that everyone is a winner? According statisticians, if you are expecting 10 guests for dinner on Christmas Day, then 15 crackers should be sufficient for everyone to win a prize. A group of statisticians arrived at this by simulating 10,000 parties, with guest numbers ranging from 2 to 50. Their results have been published in the aptly titled journal *Significance*[11].

---

[11] See: http://eu.wiley.com/WileyCDA/PressRelease/pressReleaseId-

The outcome is based on the scenario where dinner guests sit around a table, cross arms, and pull crackers with their two immediate neighbors. With this, each person has a 1 in 4 chance of winning zero crackers. A better approach is to pair up people and to have each pair pull a single cracker. This will produce the same number of winners and losers. Those who have not yet won should continue to pull crackers until only a single individual remains. That individual then pulls a cracker with themselves and everyone is a winner.

114702.html

Interim - Christmas word and math puzzle

Before moving on, here is a Christmas word puzzle for your attention!

1. Santa can see up to 25 children per hour. What is the maximum number of children he can see in 6 hours?

2. The price for each reindeer harness is $17.95. What is the price for 9 reindeer harnesses?

Challenge: If the tax rate is 14%, how much would 9 reindeer harnesses including tax cost?

3. Lana Elf measured the width of 9 Christmas presents. The total width of all the presents was 135 inches. If the presents were all the same width, how wide was each present? If all the presents had all different widths, what is one possibility for the widths of the 9 presents?

4. Peter Elf had 149 presents (not all for him) on Christmas Eve. On Christmas Day, he received 132 more presents and he gave away 128 presents. How many presents did he have to open for himself on Christmas Day?

5. Thirty elves would like to build a skating rink, so they can all use it at the same time. Santa tells them, they need at least 40 square feet for each skater, so no one will bump into each other.

If they build a rectangular rink, what are some possible dimensions (length and width) for the rink?

6. Each batch of Mrs. Crawley's toffee makes 14 pieces of toffee. If Sandy needs 8 dozen pieces of toffee, how many batches does she have to make?

7. For the Christmas dance, the dance committee needs three hours of music. Each song is an average of 3.5 minutes. How many songs do they need?

8. A candy cane factory has to order boxes for its candy canes. If they plan to make 300,000 candy canes and each box can hold 12 candy canes, how many boxes do they need to order.

9. Mr. Anderson wants to decorate six of his windows with garland. Two of the windows are 3.4 feet by 5.2 feet and the other four windows are 3.6 feet by 4.8 feet. How many feet of garland are needed for all the windows?

10. Santa needs to order boots for all of his reindeer. He has four reindeer pens with 12, 19, 14 and 16 reindeer in them. How many boots   are   needed?

The answers are over the page.

Christmas word and math puzzle answers

1. Santa can see up to 25 children per hour. What is the maximum number of children he can see in 6 hours?

150 children

2. The price for each reindeer harness is $17.95. What is the price for 9 reindeer harnesses?

$161.55

Challenge: If the tax rate is 14%, how much would 9 reindeer harnesses including tax cost?

$184.17

3. Lana Elf measured the width of 9 Christmas presents. The total width of all the presents was 135 inches. If the presents were all the same width, how wide was each present? 15 inches If all the presents had all different widths, what is one possibility for the widths of the 9 presents?

Various answers are possible here as long as they add up to 135 and they are all different amounts.

4. Peter Elf had 149 presents (not all for him) on Christmas Eve. On Christmas Day, he received 132 more presents and he gave away 128 presents. How many presents did he have to open for himself on Christmas Day?

153 presents

5. Thirty elves would like to build a skating rink, so they can all use it at the same time. Santa tells them, they need at least 40 square feet for each skater, so no one will bump into each other. If they build a rectangular rink, what are some possible dimensions (length and width) for the rink?

Various answers are possible as long as the width and the length multiply together to make 1200 sq. ft. (e.g. 20 x 60, 30 x 40, 25 x 48).

Maybe discuss the shape that would make the most sense since a 1 x 1200 ft. rink probably wouldn't be practical.

6. Each batch of Mrs. Crawley's toffee makes 14 pieces of toffee. If Sandy needs 8 dozen pieces of toffee, how many batches does she have to make?

7 batches

7. For the Christmas dance, the dance committee needs three hours of music. Each song is an average of 3.5 minutes. How many songs do they need?

180 minutes divided by 3.5 minutes or 1800 divided by 35 is probably the easiest way to do this. One of the numbers has to be converted to the other units regardless. About 52 songs.

8. A candy cane factory has to order boxes for its candy canes. If they plan to make 300,000 candy canes and each box can hold 12 candy canes, how many boxes do they need to order?

25,000 boxes

9. Mr. Anderson wants to decorate six of his windows with garland. Two of the windows are 3.4 feet by 5.2 feet and the other four windows are 3.6 feet by 4.8 feet. How many feet of garland are needed for all the windows.

Since this is a perimeter question, students also need to add the other two sides for each window. 34.4 feet + 67.2 feet = 101.6 feet of garland.

10. Santa needs to order boots for all of his reindeer. He has four reindeer pens with 12, 19, 14 and 16 reindeer in them. How many boots are needed?

61 reindeer x 4 boots per reindeer is 244 boots

5 TECHNOLOGY

Technology features heavily at Christmas, and helps create some magic. It's important, however, to make sure devices are safe.

Safety and technology

One of the common causes of accidents at Christmas is linked to lights, either from fires or people falling off roofs.

With heights, it's always best to decorate with a friend or family member so they can hold ladders and help you string the lights. It's also never a good idea to get up on the roof right after a snow or ice storm. Try and get your decorating done as early as possible before any wintery conditions make decorating slippery and unpleasant.

Now some U.K. safety facts:

In 2002, 1,000 people were estimated to have visited hospital in the UK after home accidents involving Christmas trees and 350 people after home accidents involving Christmas lights.

Fairy lights went up in smoke causing 20 fires, while Christmas trees, decorations and cards were also a fire risk and responsible for 47 house fires, leading to 20 non-fatal casualties, across the UK.

With more old-fashioned means of creating light, candles sparked around 1,000 UK house fires, resulting in 9 deaths and 388 casualties, in 2011/12.

Fairy lights and Wi-Fi

Over the Christmas period, millions of fairy lights will be installed in homes worldwide. They may appear harmless but according to Ofcom, the UK telecoms watchdog, they create problems. The regulator warns that Christmas lights can slow down Wi-Fi.

In 2015, Ofcom released a Wi-Fi Checker app. The app that is designed to let residents improve their broadband signal. In the press release about the app, the regulator made a short reference to how interference from other electronics can slow down Wi-Fi, including "Christmas fairy lights" in a list of examples. Perhaps predictably, the media has been rather more interested in this than the app itself.

So could fairy lights actually be the source of your Christmas Wi-Fi woes? It is known that electrical components can interfere with each other, particularly when a device like a Wi-Fi router is trying to get radio waves through the walls of your home. The usual sources are devices which themselves also communicate wirelessly though.

Bluetooth, wireless home security systems, wireless video game controllers and cordless phones are all notorious for interfering with routers. According to Ofcom, fairy lights should be added to the list, a claim that has risen multiple times before in the past and usually resurfaces every Christmas.

Internet service providers reportedly see a spike in complaints about Wi-Fi performance during the Christmas season. This has apparently been attributed to all the flashing lights that home routers have to put up with.

Picture 12: A tired dog caught up in the fairy lights.

Flashing fairy lights, particularly newer sets, are likely to be made from LEDs, a known source of electrical interference that can interrupt broadband signals. With so many lights around the house at Christmas time, it's therefore certainly possible that they could be responsible for a sudden drop in Wi-Fi performance.

Other possible causes should be considered too though. For most people, pulling down the lights won't be an option but some savvy placement could help with a weak Wi-Fi signal. It is recommended that cabling is kept at least 5m away from your router, if possible, giving it a space to operate in which it is free from interference.

An easy way to check how much interference your Wi-Fi is experiencing is to use an app like Ofcom's new Wi-Fi Checker. It lets you see how much data loss your router encounters as it communicates with devices on your network and can provide tips to help you optimise it.

If you're hosting a large party of visitors who are all using your Internet, any slow-down could be caused simply by this extra traffic. You could try to set bandwidth limits if your router allows you, preventing one person from using it all, or distribute connected devices between the 2.4GHz and 5GHz bands of your router, preventing them from interfering with each other.

Cheaper routers, especially ones supplied for free by ISPs, aren't always designed to handle multiple devices at once and may have slow CPUs. If you regularly use multiple devices and experience signal drop-outs — or you just want a greater broadband range in your house — consider buying a new router altogether.

One final technique you can always try when you encounter issues with your Wi-Fi is the oldest in the book: just restart your router. More often than not, this will cure any cases of devices refusing to connect and restore connection speed to normal if your network is feeling sluggish.

With a few considerations, your Wi-Fi should last you through Christmas. Just remember there are many factors that affect the quality of a wireless signal and you're almost certain to never encounter one on its own.

Interim - a Christmas word search

Santa Word Search

```
X U R S C T X B O J I W W I Y
S O W X H O Q T O C N F F I Q
L W K T I Y M T Q O A S F D Q
R F F B M S E Q A I U C Y Z B
V E K Y N S R S R H S O H L K
K Y I V E X L I U L G Q Q C C
T H C N Y U E E L U G I S K I
M D C E D S K E D X H E E Z N
J O L L Y E B U G C V E O L T
F U O J D H E H P L O D U R S
Z E U C G T D R E N K O W X Q
W F I I G Q J Z G E H W C M D
D X E L O Y P J H I W K X Z Y
N L N O R T H P O L E O S L Z
S T Y R V U D M Q A W K U Z A
```

Chimney	Rudolph
Elves	Sled
Fairies	Sleigh
Jolly	Sleigh Bells
North Pole	St. Nick
Reindeer	Toys

The Science Of Christmas: A Miscellany

6 KISS ME QUICK

Mistletoe is a plant that grows on willow and apple trees. The tradition of hanging it in the house goes back to the times of the ancient Druids. It is supposed to possess mystical powers which bring good luck to the household and wards off evil spirits. It was also used as a sign of love and friendship in Norse mythology and that's where the custom of kissing under mistletoe comes from.

The custom of kissing under mistletoe comes from England. The original custom was that a berry was picked from the sprig of mistletoe before the person could be kissed and when all the berries had gone, there could be no more kissing. The name mistletoe comes from two Anglo Saxon words 'Mistel' (which means dung) and 'tan' (which means) twig or stick. So you could translate mistletoe as 'poo on a stick'. Hmmm....

With mistletoe there may be another use other than for kissing couples. European mistletoe (*Viscum album*) is a poisonous and semi-parasitic plant that grows on a number of tree species. However, some scientists think that extracts from it might have cancer fighting properties. However, according to Lab Manager magazine, the U.S. National Cancer Institute's Physician Data Query database cautions that "most clinical studies conducted to date have had one or more major weaknesses that raise doubts about the reliability of the findings." So, more research is clearly needed[12].

---

[12] See: http://www.labmanager.com/news/2014/12/medical-mistletoe-can-the-holiday-plant-really-fight-cancer-

Picture 13: The Christmas kissing tradition...get the sprig of mistletoe ready.

In relation, cranberries are a popular Christmas accompaniment. Cranberries are a group of evergreen dwarf shrubs or trailing vines in the subgenus *Oxycoccus* of the genus *Vaccinium*. Raw cranberries have moderate levels of vitamin C, dietary fiber and the essential dietary mineral, manganese[13].

Consuming cranberry products has been associated with prevention of urinary tract infections (UTIs) for decades. However, is this popular belief a myth, or scientific fact? According to researchers in McGill University's Department of Chemical Engineering there is some truth in the bacteria-thumping properties of cranberries. Researchers have shed light on the biological mechanisms by which cranberries impart protective properties. This is by slowing down bacterial movement[14].

---

[13] See: http://nutritiondata.self.com/facts/fruits-and-fruit-juices/1875/2

[14] See: http://www.nrcresearchpress.com/doi/abs/10.1139/cjm-2012-0744#.UctRHjvqlLc

Picture 14: A heart shaped arrangement of succulent cranberries.

The Science Of Christmas: A Miscellany

7 BUGS BEHIND THE CHHRISTMAS TREE

According to insect expert Associate Professor Bjarte Jordal at the University Museum of Bergen, your beautiful Christmas tree could house up to 25,000 insects, mites, and spiders. Before looking at this, we'll briefly consider the history of the Christmas tree[15].

History of the Christmas tree

Trees have a long association with Christmas. The evergreen fir tree has traditionally been used to celebrate winter festivals (pagan and Christian) for thousands of years. Pagans used branches of it to decorate their homes during the winter solstice, as it made them think of the spring to come. The Romans used Fir Trees to decorate their temples at the festival of Saturnalia. Christians use it as a sign of everlasting life with God.

Nobody is certain when Fir trees were first used as Christmas trees. It probably began about 1,000 years ago in Northern Europe. Many early Christmas Trees seem to have been hung upside down from the ceiling using chains (hung from chandeliers/lighting hooks). The first documented use of a tree at Christmas and New Year celebrations is in town square of Riga, the capital of Latvia, in the year 1510.

---

[15] See: http://www.history.com/topics/christmas/history-of-christmas-trees

Picture 15: A Christmas tree graphic

The first Christmas Trees came to Britain sometime in the 1830s. They became very popular in 1841, when Prince Albert (Queen Victoria's German husband) had a Christmas Tree set up in Windsor Castle. Given how influential the Victorian were in re-inventing Christmas traditions, much of the modern popularity stems from this event, with middle class households copying the antics of the royal family.

Artificial trees are today found in equal number to chopped down Norwegian furs (or the equivalent). Artificial Christmas Trees started becoming popular in the early 20th century. In the Edwardian period Christmas Trees made from colored ostrich feathers were popular at 'fashionable' parties.

World records

A couple of interesting facts:

The record for the most Christmas trees chopped down in two minutes is 27 and belongs to Erin Lavoie from the USA. She set the record on 19th December 2008 on the set of Guinness World Records: Die GroBten Weltrekorde in Germany.

The tallest artificial Christmas tree was 52m (170.6ft) high and was covered in green PVC leaves!. It was called the 'Peace Tree' and was designed by Grupo Sonae Distribuição Brasil and was displayed in Moinhos de Vento Park, Porto Alegre, Brazil from 1st December 2001 until 6th January 2002.

What's lurking behind your tree?

Quite a bit according to Associate Professor Bjarte Jordal at the University Museum of Bergen[16]. In an interview, Jordal explains:

"In research on Christmas trees there have been found as many as 25,000 individual creep in some of the trees. If you pound the tree on a white cloth before you throw it out after Christmas, you will discover quite a number of small bugs."

How do these creeps end up in the Christmas tree?

"They go to sleep for the winter, or hibernate to use the technical term. They usually empty their bodies of fluids and produce a chilled liquid and are completely inactive. But they reawaken when the tree is brought into the heat of the living room. It's all down to stimulus. Upon feeling the heat and awakened by the light, they believe that springtime has arrived and spring back to life."

So do they go about wandering around the living room or what?

"No, I believe they stay in the tree. Both the Christmas tree and

---

[16] See: http://www.uib.no/news/nyheter/2012/12/bugs-in-the-christmas-tree

the house itself will be very dry. Also, most insects don't live off the tree, only in it. As they cannot feed on the limited plants found in most households, the bugs will quickly dry out and die. These insects and bugs do not constitute any risk or danger to people or furniture. And if anyone is worried about allergic reactions, I don't think there's any danger of that. But obviously, should there be an extreme number of mites in a tree people with severe allergies may react to this."

Are there a fixed number of bugs in each and every Christmas tree?

"This varies a lot. Some of it is down to pure coincidence and some of it is down to what type of tree it is. Trees chopped in your own backwoods will contain more bugs than firs and other trees that have been farmed for use as Christmas trees will contain fewer creeps. There is particularly much in Norwegian Pine, whereas Juniper shrub has a fauna of its own."

Can you spot the little beasts on the tree?

"No, they are good at hiding and are invisible to the human eye, although one certainly should be able to spot the odd spider. To get a proper look, you will have to get out a clean, white sheet and shake the tree."

What about the tabloid media's favourite arachnids – the ticks? Can they be found in our Christmas trees as well?

"There may very well be, and the Norwegian Institute of Public Health has actually looked into this. Their research suggests that there are three reports every Christmas in Norway of ticks found in Christmas trees. What usually happens is that the family dog has gone to rest under the tree and has incurred ticks. But the overall chances of tick bites are minimal. Also, the dog need not be allowed to rest under the tree. And the ticks are usually in sleep mode when the tree is brought into the house and dead by the time the tree leaves the house after Christmas. So, as I said, the risk is minimal."

But even if there is seemingly little danger or nuisance to expect from these creepy-crawlies, what should people be conscious of to minimize the number of bugs in the Christmas tree?

"I would recommend that you get a locally grown hardwood tree, as this is most likely to have a limited fauna. But you should by no means clean or flush the tree free of bugs, as this will damage the tree. Anyway, there is nothing to fear. You need to take into consideration that there are plenty of insects and bugs in potted plants that are regular features in most households. As we all know, these attract plenty of flies. It's no different with Christmas trees."

Do you think that people are aware that the Christmas tree they bring into the house is full of little bugs?

"Probably not. After all, these little bugs are invisible to the human eye. I believe there is a trend in people not being particularly knowledgeable about nature. But when you bring a tree into the comfort of your living room, the tree carries a part of nature with it. Yet at the same time people tend to remove themselves more and more from nature."

The Science Of Christmas: A Miscellany

8 SAFETY

While safety has been touched on a little in relation to technology, it's worth reminding ourselves about safety over Christmas. Many of these are rooted in science (and commonsense).

Christmas safety tips:

1. Make sure you buy children's gifts for the correct age group and from reputable sources that comply with standards.
2. Remember to buy batteries for toys that need them - that way you won't be tempted to remove batteries from smoke alarms.
3. Look out for small items that could pose a choking hazard to young children, including parts that have fallen off toys or from Christmas trees, button batteries and burst balloons.
4. Keep decorations and cards away from fires and other heat sources such as light fittings. Don't leave burning candles unattended, make sure you put them out before going to bed and do not put candles on Christmas trees.
5. If you have old Christmas lights, seriously consider buying new ones, which will meet much higher safety standards, keep the lights switched off until the Christmas tree is decorated, don't let children play with lights (some have swallowed the bulbs), and remember to switch off the lights when going out of the house or going to bed.
6. Remember, Christmas novelties are not toys, even if they resemble them, and they do not have to comply with toy safety regulations. Give careful thought to where you display them, for example, place them high up on Christmas trees where they are out of the reach of young hands.
7. Give yourself enough time to prepare and cook Christmas dinner to avoid hot fat, boiling water and sharp knife accidents that come from rushing, and keep anyone not helping with dinner out of the kitchen. Wipe up any spills

quickly.

8. Have scissors handy to open packaging, so you're not tempted to use a knife, and have screwdrivers at the ready to assemble toys.

9. Beware of trailing cables and wires in the rush to connect new gadgets and appliances, and always read instructions

10. Falls are the most common accidents so try to keep clutter to a minimum. Make sure stairs are well-lit and free from obstacles, especially if you have guests.

11. Plan New Year fireworks parties well in advance.

12. Do not drink and drive, and plan long journeys so you won't be driving tired.

13. Finding a coin in your pudding on Christmas day - it's a tradition that's lasted for more than 500 years and is said to grant you a good luck wish for the coming year. Just make sure you don't swallow it.

9 CHRISTMAS SCIENCE EXPERIEMNTS

Experiments for children (or even adults) can be fun around Christmastime and be educational. Here are a selection of five different things to try.

A. Growing crystals

A crystal is a solid material with a naturally geometrically regular form. Some take millions of years to form, such as diamonds. The crystals we made above take just a few days.

Most minerals dissolved in water will form crystals given enough time and space. The shape of the crystal formed depends on the mineral's molecule shape.

In the case of our sugar crystals there are two process at work.

Evaporation – the water evaporates slowly meaning the solution becomes more saturated, so the sugar molecules come out of solution and collect on the string/wire or stick.

Precipitation – the solution we made was very concentrated which means there was too much solute to remain dissolved in the water, therefore it starts to precipitate.

<u>What you need</u>

- 3 cups of caster sugar

- 1 cup of water

- A lolly stick or circle of wire. Or some string.

- A Jar,

- Stikcy-tape.

- Some sparkles and/or food coloring(optional)

Instructions

Dissolve the sugar in the water, as soon as it is dissolved remove the heat and leave to cool a little bit. If you can get a bit more sugar to dissolve then thats a good thing. We want a saturated solution.

Pour the solution into a glass jar and suspend the lolly stick, we used some sellotape to hold it in place. Don't let it touch the bottom or the sides. Alternatively you could tie some string to a pencil and rest the pencil on top of the jar with the string hanging in the jar.

You should see crystals start to form after a few days.

Picture 16: Home-grown crystals

B. Learning about filtration...and making spiced apple

A filter is a porous material which a liquid can be passed though to separate the liquid from solids suspended in it. In this activity we used a muslin cloth to filter out the big bits of apple, leaving just the juice. To make the juice less cloudy we could have filtered it through something with smaller holes, to separate more of the solid bits from the liquid.

To illustrate filtration to children, why not make some spiced apple beverage?

Ingredients:

- Apples or apple juice

- Spices – cinnamon sticks, cloves, nutmeg and cardamon

- You will need a big pan, a muslin cloth and something to bash the spices with.

Instructions:

- Core about 10 apples, chop them up and then blend them with a hand blender.
- Lay out a muslin and we pile the apple pulp in the center.
- Gather up the corners of the muslin and then squeeze every last drop of apple juice through the filter. All but the very smallest parts of apple and juice should be able to pass through the cloth's weave.
- With a second bit of muslin add some spices. Cloves, Star A nice, nutmeg, cinnamon, cardamom. Bash them in a mortar with a pestle and then tie up the muslin to make a little bag.
- Then add the spice and apple juice to a saucepan and heat gently for about 20 minutes.

Picture 17: Delights of a good spiced apple drink

C. Firework in a glass

The science bit: Oil and water do not mix! Also Oil is less dense than water (meaning there is less of it in the same volume) and therefore floats on top of water in a nice layer.

The food coloring we used was water based and therefore does not mix with the oil, instead it sinks through the oil into the water below.

Since the addition of the coloring makes the food coloring heavier than the water it sinks to the bottom leaving trails (resembling fireworks) as some of the color diffuses into the water.

The experiment

Before you begin, you'll need:

- A tall glass and a smaller glass

- Warm water

- Oil

- Food coloring

The process:

- Fill a tall glass with warm water almost to the top.
- Pour a couple of tablespoons of oil into another glass and add a few drops of food coloring.
- Mix up the oil and food coloring.
- Pour the food coloring and water mixture into the warm water and watch the fireworks.

D. The physics of cooking

Making peppermint creams can be help to illustrate different states of matter.

This experiment is a great way to demonstrate the process of changing state from solid to liquid and back again.

SOLID —> LIQUID = MELTING

LIQUID —> GAS = EVAPORATING/BOILING

GAS —> LIQUID = CONDENSING

LIQUID —> SOLID = COOLING/FREEZING

When the chocolate was heated it changed from solid to liquid, which is an example of melting and when we move back from a liquid into a solid it is an example of cooling/freezing.

The reason this happens is because when you provide heat the particles that make up the solid are given energy which cause them to vibrate and then break the bonds holding them together. As they cool they lose this energy and so forms bonds again but not in the same shape…this is why we can mould chocolate and other substance.

The sweet making process

Ingredients

- 225g Icing Sugar

- 115g condensed milk

- A few drops peppermint extract

- 55g plain chocolate melted

Instructions

- Mix the icing sugar with the condensed milk and knead until you have a smooth consistency.
- Roll out and use cookie cutters to shape.
- Leave for about 3 hours to harden and then coat with the melted chocolate.

Picture 18: Make your own tasty peppermint creams and learn about states of matter.

E. Make a Christmas-time glitter globe

In this science experiment we will mix molecules to make a glitter globe (i.e. a "snow globe"). We will combine rubbing alcohol, vegetable oil, and a few other tiny, shiny things to make a cool science toy.

The science bit: Due to the difference in densities, the oil floats on top of the water. In this case however, the oil sinks below the alcohol. The interaction between two oil molecules is weaker than the interaction between two alcohol molecules and hence, they don't mix well.

Alcohols are polar molecules which attract each other with strong permanent dipole forces. Oil on the other hand, is non-polar and so only attracts other oil molecules with temporary dipole forces that are much weaker. The oil ends up sinking to the bottom of the bottle because the oil is more dense than alcohol.

How to make a glitter globe

Supplies:

- Rubbing Alcohol (isopropyl),
- Vegetable oil or baby oil,
- Clear plastic bottle or glass container,
- Small beads (and/or sequins, glitter, or anything else tiny and shiny),
- Food coloring,
- Clear tape.

Instructions:

1. Fill a clear plastic or glass bottle 1/4 full of rubbing alcohol.

2. Add one drop of food coloring if you want to give the liquid mix some color (and make it easier to differentiate the alcohol layer from the layer of oil that we will add

next).Note: if you want to use your snow globe for decorative purposes, skip the food coloring altogether.

3. Fill the remainder of the bottle with baby or vegetable oil (the oil will sit on top of the alcohol). Leave a bit of room in the lid area for the tiny, shiny stuff.

4. Add glitter, sequins, small beads or any other tiny, shiny objects.

5. Fill the remainder of the bottle, all the way to the rim, with baby oil.

6. Screw the lid on tightly. Tape it so it does not come loose (or apply a little glue to the inside of the lid before twisting in on).Note: If you are using a wide-mouth jar for your bottle, you can also glue some sort of decoration (e.g. church, snowman, Christmas tree, plastic deer, etc.) to the inside of the lid before putting the lid on the jar.

7. Gently shake the bottle to mix the alcohol and oil together. The mixture will turn a milky color and the shiny stuff will float and spin in the mixture.

8. Allow the mixture to settle. The oil will separate from the alcohol after about 5 minutes.

Picture 19: A home-made snow globe

10 SCIENTISTS FOR CHRISTMAS?

Which scientist could be used to represent Christmas, offering a grownup figurehead for scientists in place of Santa? A few years ago, journalist Dean Burnett drew up a list of contenders.

• *Albert Einstein*: Already has the elaborate white hair, and travelling to every house in one night with a sack which clearly has fluctuating mass suggests some sort of relativistic physics is occurring.

• *Charles Darwin*: Had a big white beard, so not much of an image shift needed. Gave the gift of understanding to mankind (sort of). Could also be useful for the traditional belief v evidence plot of classic Christmas films.

• *Rosalind Franklin*: A generous soul who gave much to others, but never really got the credit she deserved for her efforts. She did actually exist, as well. Taken together, she's essentially the exact opposite of Santa Clause, so maybe not such a good choice after all.

• *Alan Moore*: A big beard, does magic, could come up with his own brilliant and harrowing origin story.

• *Ada Lovelace*: Arguably the first computer programmer, so an appropriate modern-day choice given that every gift seems to be a gadget of some sort. Although given her admiration for her father, she may struggle with the whole 'naughty/nice' system.

• *Patrick Moore*: Another Moore. The late, great Patrick Moore could be celebrated every year for his gift of enthusiasm and celebration of the universe, as well as his support for child-orientated things like video games. He may have had some controversial opinions, but an older, friendly-relative suddenly spouting shocking opinions is very much in-keeping with the spirit of festive family gatherings.

• *David Attenborough*: Just because.

11 WHY DO I FEEL ILL?

Christmas is commonly a time for over-indulging. Living to the excess can sometimes make you feel ill. Why is this? Here are some scientific explanations.

Seasonal affective disorder

Many poeple suffer from seasonal affective disorder – severe changes in mood with the onset of winter. Those affected are thought to have low levels of serotonin and melatonin, which means they require much more daylight.

When light hits the retina at the back of the eye, electrical signals are sent to the hypothalamus. This is the part of the brain that controls sleep, appetite, body temperature, sex drive and mood. When there's not enough light getting through, these functions begin to slow down.

Over-eating

Gastric indulgence is pretty much obligatory at Christmas – and indigestion is a familiar consequence.

But over-eating also has a far more sinister effect. It stimulates a usually dormant pathway between the hypothalamus and the immune system. This results in an excessive immune response and leads to low grade inflammation throughout the body. And that's why you often feel unwell after eating too much.

Over-eating for long periods of time leads to chronic inflammation, which can contribute to Type 2 diabetes and heart disease.

Too much alcohol

Infrequent heavy drinking sessions at Christmas and New Year are unlikely to have a serious health effect, but you may experience short-term symptoms such as feeling anxious or irritable when you stop. Abrupt cessation of alcohol use leads to brain hyperexcitability due to the sudden "release" effect on receptors which had been inhibited by the alcohol.

Too much television

People are reluctant to accept that TV has any effect on them, but doing anything for that long will have some short-term neurological consequences.

Television encourages low alpha waves (brainwaves in the frequency range 8-12 Hz) in the brain. These brainwaves are associated with relaxation, but also suggestibility – something advertisers look to capitalize on by making your receptive brain associate their brands with positive emotions.

If you spend many hours with your brain operating in the low-alpha state, this can result in attention-span issues and an inability to concentrate when you return to work.

Family stress

Christmas is a time for family. For many, that is not an unmitigated blessing. Any form of stress leads to the release of adrenaline and cortisol – an automatic response that has been in our genes since we were hunter gatherers.

But there is one part of a brain that is especially vulnerable to increases in cortisol and that's the hippocampus. As a result, when you're stressed you may find your ability to multi-task and remember things is impaired. Not hugely helpful if you're making the meal.

Exercising and getting plenty of sleep are the best ways to

counteract increased cortisol levels.

Exercise stimulates the growth and repair of cells in the hippocampus, and induces a more positive mood, making you less susceptible to stress and reducing the amount of cortisol released. Plus it provides the perfect excuse to run away from your relatives.

12 CHRISTMAS COLORS

The meaning of colors may not be scientific in the sense of natural sciences, but there are psychological implications. With this in mind, what do the colors most commonly associated with Christmas mean?

Green

Evergreen plants, like Holly, Ivy and Mistletoe have been used for thousands of years to decorate and brighten up buildings during the long dark winter. They also reminded people that spring would come and that winter wouldn't go on forever.

The Romans would exchange evergreen branches during January as a sign of good luck. The ancient Egyptians used to bring palm branches into their houses during the mid winter festivals.

In many parts of Europe during the middle ages, Paradise plays were performed, often on Christmas Eve. They told Bible stories to people who couldn't read. The 'Paradise Tree' in the garden of Eden in the play was normally a pine tree with red apples tied to it.

Now the most common use of green at Christmas are Christmas Trees.

Red

As mentioned above, an early use of red at Christmas were the apples on the paradise tree. They represented the fall of Adam in the plays.

Red is also the color of Holly berries, which is said to represent the blood of Jesus when he died on the cross.

Red is also the color of Bishops robes. These would have been worn by St. Nicholas and this could be one reason why Santa is commonly dressed in red.

Gold

Gold is the color of the Sun and light - both very important in the dark winter. And both red and gold are the colors of fire that you need to keep you warm.

Gold was also one of the presents brought to the baby Jesus by one of the wise men and traditionally it's the color used to show the star that the wise men followed.

Silver is sometimes used instead of (or with) gold. But gold is a 'warmer' color.

White

White is often associated with purity and peace in western cultures. The snow of winter is also very white!

White paper wafers were also sometimes used to decorate paradise trees. The wafers represented the bread eaten during Christian Communion or Mass, when Christians remember that Jesus died for them.

White is used by most churches as the color of Christmas, when the altar is covered with a white cloth (in the Russian Orthodox Church Gold is used for Christmas).

Blue

The color blue is often associated with Mary, the mother of Jesus. In medieval times blue dye and paint was more expensive than gold. So it would only be worn by wealthy people. Blue can also represent the color of the sky and heaven.

During Advent, purple and sometimes blue is used in most churches for the color of the altar cloth (in the Russian Orthodox Church red is used for advent).

13 SNOW AND ICE

Snow, Ice and Christmas often go together, although why it should is a bit strange. The reason that we think of Snow and Ice at Christmas is portably down to the Victorians. Although Christmas was taken over from the Pagan winter solstice festivals in Europe, it was the Victorians who produced this idea of the 'traditional' Christmas in Europe and the USA.

At the start of the Victorian era, (1837) Britain was in a mini ice age that was from about 1550 to 1850.

With this in mind, here are some snow and ice science facts.

How Snow Forms

Once snow crystals form in the atmosphere, they grow by absorbing surrounding water droplets. The snowflakes we end up seeing on the ground are an accumulation of these ice crystals. This magnified image of snow crystals was captured by a low-temperature scanning electron microscope (SEM). The pseudo colors commonly found in SEM images are computer generated, and in this case highlight the different flake formations.

Whether winter storms produce snow relies heavily on temperature, but not necessarily the temperature we feel here on the ground. Snow forms when the atmospheric temperature is at or below freezing (0 degrees Celsius or 32 degrees Fahrenheit) and there is a minimum amount of moisture in the air. If the ground temperature is at or below freezing, the snow will reach the ground. However, the snow can still reach the ground when the ground temperature is above freezing if the conditions are just right. In this case, snowflakes will begin to melt as they reach this higher temperature layer; the melting creates evaporative cooling which cools the air immediately around the snowflake. This cooling retards melting. As a general rule, though, snow will not form if the ground temperature is at least 5 degrees Celsius (41 degrees Fahrenheit).

While it can be too warm to snow, it cannot be too cold to snow. Snow can occur even at incredibly low temperatures as long as there is some source of moisture and some way to lift or cool the air. It is true, however, that most heavy snowfalls occur when there is relatively warm air near the ground—typically -9 degrees Celsius (15 degrees Fahrenheit) or warmer—since warmer air can hold more water vapor.

Because snow formation requires moisture, very cold but very dry areas may rarely receive snow. Antarctica's Dry Valleys, for instance, form the largest ice-free portion of the continent. The Dry Valleys are quite cold but have very low humidity, and strong winds help wick any remaining moisture from the air. As a result, this extremely cold region receives little snow.

Snow on the ground

The character of the snow surface after a snowfall depends on the original form of the crystals and on the weather conditions present when the snow fell. For example, when a snowfall is accompanied by strong winds, the snow crystals are broken into smaller fragments that can become more densely packed. After a snowfall, snow may melt or evaporate, or it may persist for long periods. If snow persists on the ground, the texture, size, and shape of individual grains will change even while the snow temperature remains below freezing, or they may melt and refreeze over time, and will eventually become compressed by subsequent snowfalls.

Over the winter season, the snowpack typically accumulates and develops a complex layered structure made up of a variety of snow grains, reflecting the weather and climate conditions prevailing at the time of deposition as well as changes within the snow cover over time.

Picture 20: A classic snowflake

How big can snowflakes get?

Snowflakes assume any of a number of shapes, depending on temperature and other conditions. Snowflakes are accumulations of many snow crystals. Most snowflakes are less than 1.3 centimeters (0.5 inches) across. Under certain conditions, usually requiring near-freezing temperatures, light winds, and unstable atmospheric conditions, much larger and irregular flakes can form, nearing 5 centimeters (2 inches) across. No routine measure of snowflake dimensions are taken, so the exact size is not known.

Melting snow

When you add salt to water, you introduce dissolved foreign particles into the water. The freezing point of water becomes lower as more particles are added until the point where the salt stops dissolving. For a solution of table salt (sodium chloride, NaCl) in water, this temperature is -21°C (-6°F) under controlled lab conditions. In the real world, on a real sidewalk, sodium chloride can melt ice only down to about -9°C (15°F).

Make your own snowflake

Grow a borax crystal snowflake, color it blue if you like, and enjoy the sparkle all year long.

Borax Crystal Snowflake Materials

- string
- wide mouth jar (pint)
- white pipe cleaners
- borax (see tips)
- pencil
- boiling water
- blue food coloring (opt.)
- scissors

Time Required: Overnight.

For the process:

1. The first step of making borax crystal snowflakes is to make the snowflake shape. Cut a pipe cleaner into three equal sections.

2. Twist the sections together at their centers to form a six-sided snowflake shape. Don't worry if an end isn't even, just trim to get the desired shape. The snowflake should fit inside the jar.

3. Tie the string to the end of one of the snowflake arms. Tie the other end of the string to the pencil. You want the length to be such that the pencil hangs the snowflake into the jar.

4. Fill the widemouth pint jar with boiling water.

5. Add borax one tablespoon at a time to the boiling water, stirring to dissolve after each addition. The amount used is 3 tablespoons borax per cup of water. It is okay if some undissolved borax settles to the bottom of the jar.

6. If desired, you may tint the mixture with food color.

7. Hang the pipe cleaner snowflake into the jar so that the pencil rests on top of the jar and the snowflake is completely covered with liquid and hangs freely (not touching the bottom of the jar).

8. Allow the jar to sit in an undisturbed location overnight.

Tips for Success

o Borax is available at grocery stores in the laundry soap section, such as 20 Mule Team Borax Laundry Booster. Do not use Boraxo soap.

o Because boiling water is used and because borax isn't intended for eating, adult supervision is recommended for this project.

o If you can't find borax, you can use sugar or salt (may take longer to grow the crystals, so be patient). Add sugar or salt to the boiling water until it stops dissolving. Ideally you want no crystals at the bottom of the jar.

Make fake snow

1. There are a couple of ways to get the ingredient necessary to make fake polymer snow. You can purchase the fake snow or you can harvest sodium polyacrylate from common household sources. You can find sodium polyacrylate inside disposable diapers or as crystals in a garden center, used to help keep soil moist.

2. All you need to do to make this type of fake snow is add water to the sodium polyacrylate. Add some water, mix the gel. Add more water until you have the desired amount of wetness. The gel will not dissolve. It's just a matter of how 'slushy' you want your snow.

3. Sodium polyacrylate 'snow' feels cool to the touch because it is mainly water. If you want to add more realism to the fake snow, you can refrigerate or freeze it. The gel will not melt. If it dries out, you can rehydrate it by adding water.

Bonus Chapter

14 SPOOKY! THE SCIENCE BEHIND HALLOWEEN

Let's begin with getting scared. Many people will watch a scary movie on October 31 and a reaction to a haunted house or creepy slasher naturally raises the goosebumps. Despite seeking fear intentionally, in most cases it is a natural reaction and one very important to a human's survival instincts: the heart races, pupils dilate, we breathe heavy, sometimes the stomach tightens, even dizziness. Talking to CBS about this issue, psychologist Dr. Elizabeth Gordon notes: "Fear is there to protect us. People fear things that would have been dangerous as we evolved. A lot of things we see at Halloween play on natural fears we have to things like spiders."[17]

Picture 21: A Brown recluse Spider

---

[17] See: http://philadelphia.cbslocal.com/2015/10/30/the-science-behind-fear/

Next, fangs. This time, not vampires but deer. Did you know deer could have fangs?

The deer in question is the Kashmir musk deer. It was recently spotted in 2014 exhibiting pointed teeth, the first recorded incident since 1948. The musk deer is classified as an endangered species on the International Union for Conservation of Nature's Red List.

Picture 22: Kashmir musk deer

Moving on to some of the classic monsters. With werewolves there is a genetic condition that leads to some people having excessive body hair. It is possible that, in years gone by, people with this condition were mistaken for half-wolf, half-human creatures. Even today the condition - hypertrichosis - is sometimes called "werewolf syndrome."

Picture 23: A boy with hypertrichosis

How about zombies? The story behind the walking dead seems to date back to people falling under trances (either natural or artificially induced through mind-altering substances.) For instance, an ethnobotanist investigating the claims in Haiti found a toxic drug that could actually induce a zombie-style catatonic state[18].

Beyond people there is the 'zombie wasp.' Here the wasp *Dinocampus coccinellae* takes control of a ladybug that carries an egg the wasp lays in its abdomen[19].

---

[18] See: http://mysteriousuniverse.org/2014/08/the-mysterious-real-zombies-of-haiti/

[19] See: http://phys.org/news/2015-08-wasp-masters-web-building-zombie-slave.html

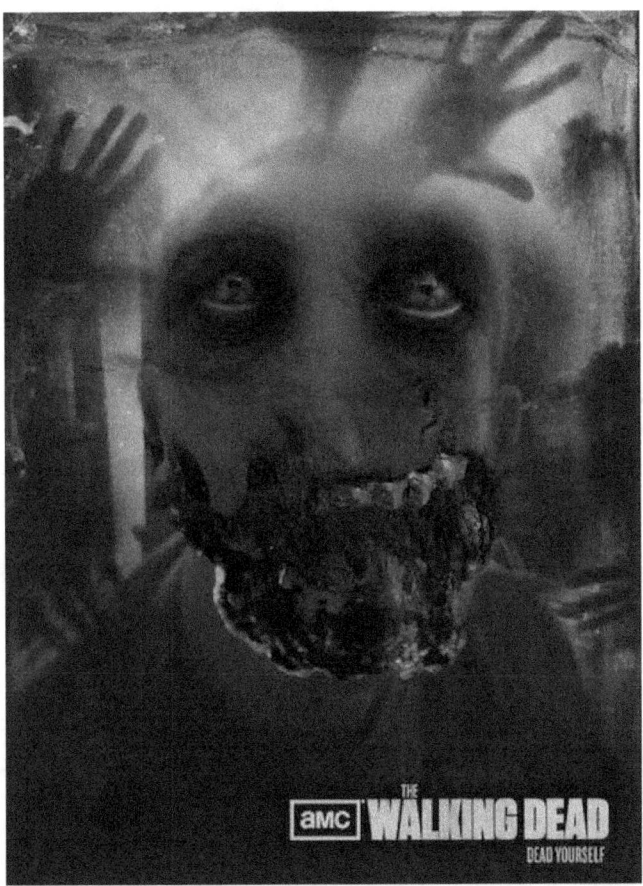

Picture 24: The author of this book transformed into a zombie using trick photography.

Going back to fear, if you were afraid of spiders would you be prepared to have a chunk of your brain removed? Recently a 44 year-old businessman, who suffered from arachnophobia, started suffering with seizures. The only way to deal with this was for him to undergo surgery to remove part of his amygdala, which is buried deep within the brain. The operation was successful and when he recovered his fear of spiders had gone[20].

---

[20] See: https://www.washingtonpost.com/news/speaking-of-science/wp/2014/10/31/man-is-cured-of-arachnophobia-by-losing-a-chunk-of-

If you want to try some scientific special effects, like smoky fingers, there are several websites outlining experiments (not for children!) If you want to create a puff of smoke from your fingers, this can be done with the help of some glycerol. Here a small quantity of glycerol (or the liquid added to fog machines) can be rubbed on the middle finger and thumb. The heat generated from this activity produces a small puff of magical smoke[21].

Finally, one way to really scare someone is with a scream. You want this to be high-pitched, right? Perhaps not according to David Poeppel, director of the Max Planck Institute for Empirical Aesthetics in Germany. What makes a scary scream is a particular modulation. This was found by subjecting a wide range of subjects, The Washington Post reports, to a range of different screams (some taken from horror films.) The most frightening are those at a modulation rate between 4 and 5 hertz. The singer Tom Waits was cited as often coming closest to the vocal range.

---

his-brain/

[21] See: http://www.teachhub.com/top-five-halloween-science-tricks

ABOUT THE AUTHOR

Tim Sandle is a science writer and journalist. By profession he is a pharmaceutical microbiologist and runs a popular microbiology website: http://www.pharmamicroresources.com/

He lives in the U.K.

www.ingramcontent.com/pod-product-compliance
Lightning Source LLC
Chambersburg PA
CBHW051341170526
45166CB00002B/900